INTERMEDIATE GEOBOARD ACTIVITY BOOK

(Grades 4–6)

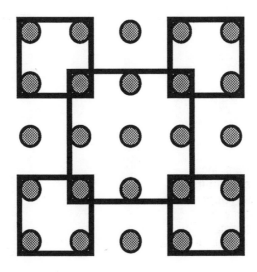

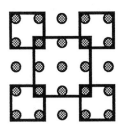

Introduction

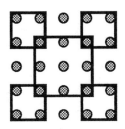

The *Intermediate Geoboard Activity Book* is a resource book of reproducible blackline masters designed to help teachers make the best use of the geoboard. The geoboard can be used to make learning the applications of geometry easier by providing hands-on activities for students.

In this book, teachers in the intermediate grades will find activities that focus on topics in plane geometry and measurement. Students will explore and internalize key geometric manipulations such as transformations and rotations, as well as geometric concepts and applications for polygons and quadrilaterals.

Each of the seven sections in the *Intermediate Geoboard Activity Book* focuses on a related group of activities. The blackline masters build sequentially on students' prior learning and involve copying, manipulating or designing on the geoboard. As students learn geometry via the manipulation, their visual thinking and spatial reasoning skills will improve. Students will also refine problem-solving strategies, particularly in terms of guessing, checking and revising their work.

Teaching suggestions are outlined at the beginning of each section:

Getting Started	Pre-learning activities to model concepts and show why they are important.
Using the Worksheets	Pacing and incremental learning hints.
Practice	Extension activities to put the concepts into practice.
Wrap-up	Checks for understanding and use of terms.

One of the most fascinating aspects of working with geoboards is the creativity which they promote in students. It is hoped that the activities in this book will inspire students to pose problems of their own or find new areas to investigate.

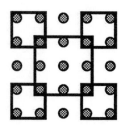
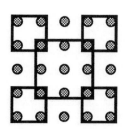

Table of Contents

Section A: Transformations

Section B: Perimeter

Section C: Area

Section D: Polygons

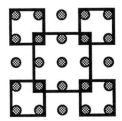

Table of Contents

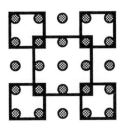

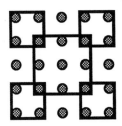

Teacher's Notes

Section A
Transformations

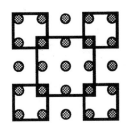

Getting Started

Use your geoboard to show students various figures to copy. First, model a transformation called a *slide* in which the entire figure is moved in any one direction. Then model a transformation called a *reflection* in which the figure is flipped over a line of symmetry. Help students discover why in each transformation their figures match exactly. Ask volunteers to explain why the two figures are congruent. **Note:** One other transformation—*rotation*—will be treated later in Section G, Rotation and Design.

Using the Worksheets

	Copying Shapes	2
	Sliding One Space Up or Down	3
	Sliding Two Spaces Up or Down	4
	Sliding Two Spaces Right or Left	5
Use the worksheets	Sliding Up/Down and Right/Left	6
in the order shown	Congruent or Not?	7
at the right.	Making Congruent Figures	8
	More Congruent Figures	9
	Flips	10
	Finding Lines of Symmetry	11
	Similar Figures	12

As you progress through the worksheets, have students begin by copying the figures exactly before performing the transformations. Ask them to leave the starting figure in place while making the particular transformation with a different colored geoband. Students should be able to accomplish the transformation by moving the geoband one corner at a time. Later, have students move the entire figure. Instruct them to mark the line of symmetry of a flip with another colored geoband.

Practice

Have students work in pairs with their geoboards. Students take turns making slides and flips from their partner's starting figure and directions. Remind students to first determine whether there is enough room for the transformation. Extend the activity by showing students some transformations that combine a flip followed by a slide. For each procedure encourage students to tell how the transformation is made.

Wrap-up

Ask students to show specific slides and flips. Have them identify the line of symmetry for each flip.

Copy each shape on your geoboard.

1.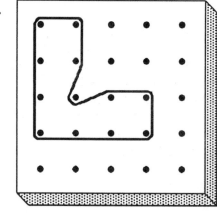

2.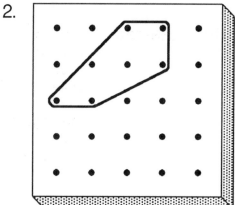

3.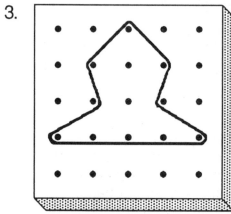

4.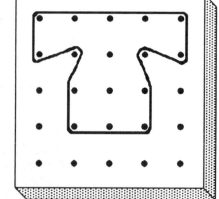

5.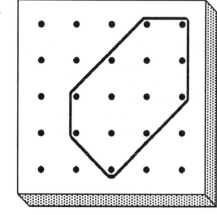

6.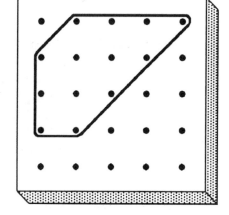

Name

Slide each shape one space down.

1.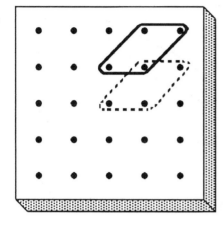

2.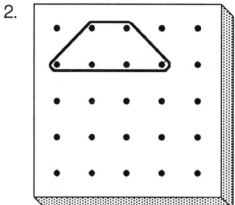

3.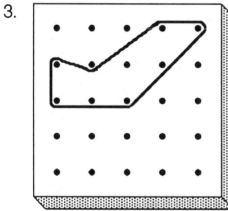

4.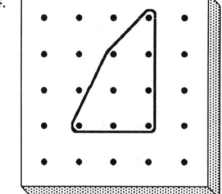

Slide each shape one space up.

5.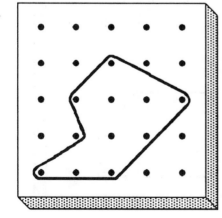

6.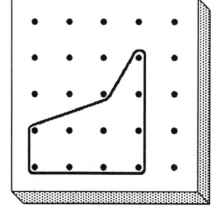

Slide each shape two spaces down.

1.

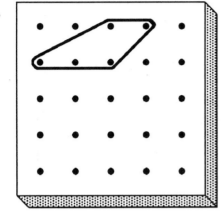

2.

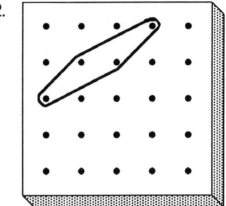

3.

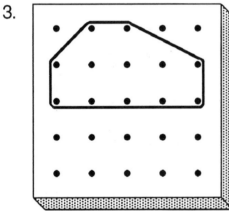

4.

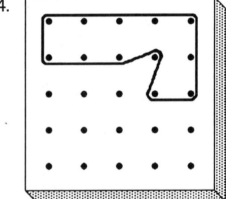

Slide each shape two spaces up.

5.

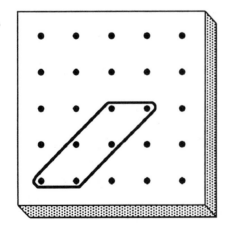

6.

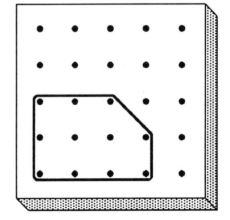

Slide each shape two spaces to the right.

1.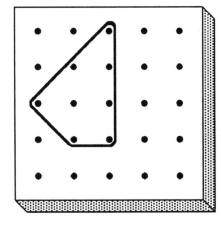

2.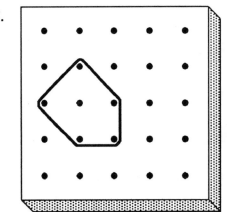

3.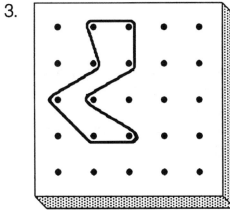

4.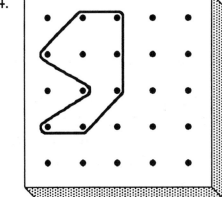

Slide each shape two spaces to the left.

5.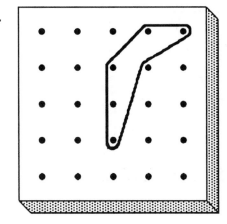

6.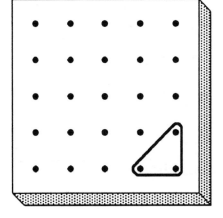

Slide the figures below two spaces down and then two spaces to the right.

1.

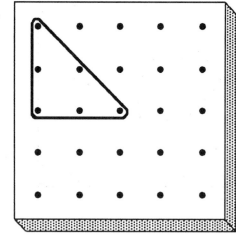

2.

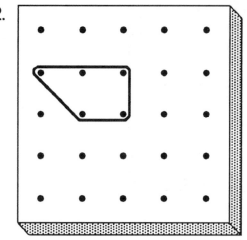

3. Kate has two running areas, A and B, in her yard for exercising her dogs. On a separate sheet, show by sliding A and B how she can rearrange the running areas.

4. The Best Ceramic company ships 2 pieces of sculpture that have flat bottoms. On your separate sheet, show by sliding how 4 C pieces and 4 D pieces can fit into a box.

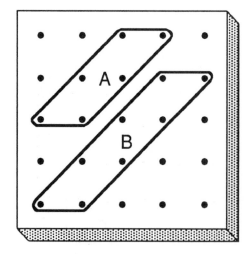

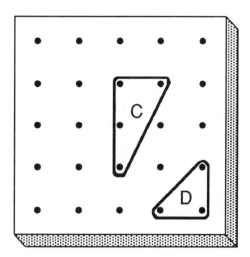

Name

Figures with the same shape and size are congruent.
Copy each pair of shapes on your geoboard. Look for congruent figure pairs.

1.

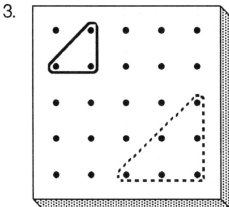

2.

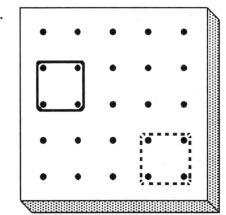

3.

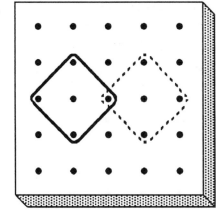

4.

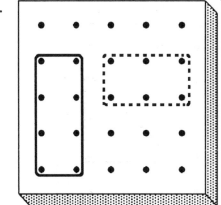

5.

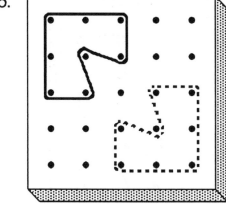

6. List the numbers of those geoboards which show congruent figures.

___ ___ ___

7. List those which do not.

___ ___ ___

Make each shape. Using another geoband, slide the shape to make
a congruent figure.

1. Two spaces right

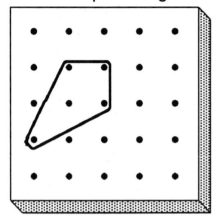

2. One space down

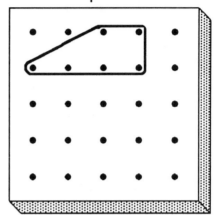

3. Two spaces left

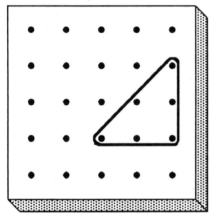

4. Three spaces up

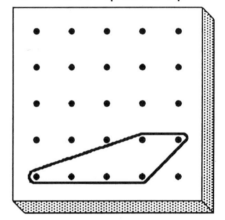

5. One space right

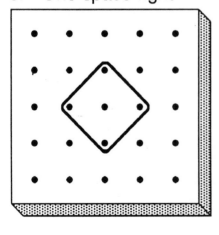

6. One space left

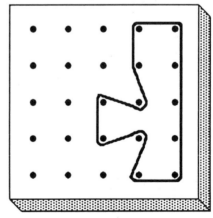

Name

Make each shape. Using another geoband, slide the shape to make
a congruent figure.

1. Two left, one down

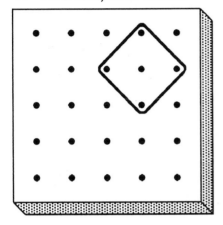

2. One right, two up

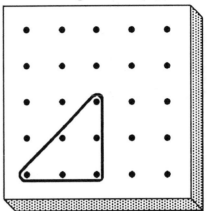

3. One down, one right

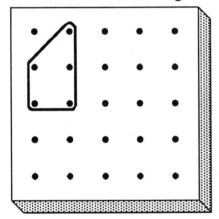

4. Two up, one left

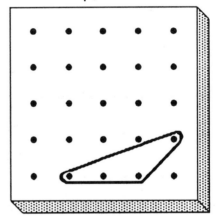

5. Two down, three right

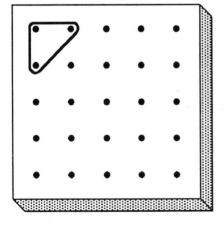

6. Three up, one right

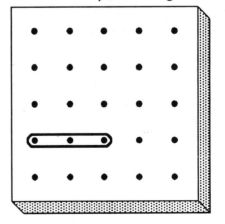

On another sheet, slide or flip each of these figures
so that 4 of them fill up the page with no overlap.

1.
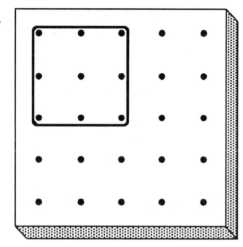

2.
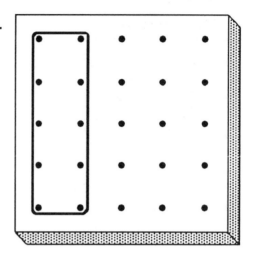

3.
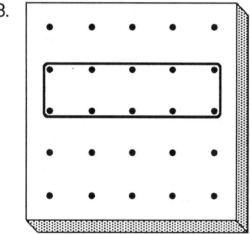

4.
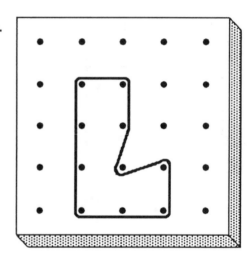

5. Use another sheet. Fold the sheet in half. Draw a line on
   the fold. Then draw a triangle on one side of the line.

6. Flip the triangle across the line. Is the second triangle
   congruent to the first triangle? _____

Name

You can show lines of symmetry with a square.
Use two paper squares of any size.

1.   Fold one square diagonally to show two congruent triangles. Draw the line
     of symmetry on the fold.

2.   Fold the other square in a different way. Draw a line of symmetry.

3.   Draw a triangle on one side of the line of symmetry.
     Flip the triangle across the line. Trace the new, congruent triangle.

Find a line of symmetry for each shape.

4.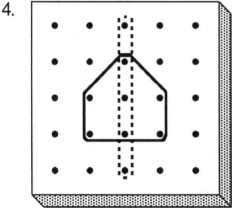

5.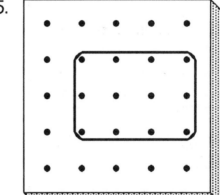

6.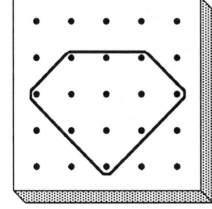

7.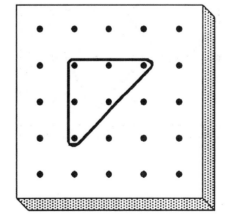

Figures with the same shape are similar.
They may be different sizes.

Circle *yes* for similar or *no* for not similar for each pair
of shapes shown below.

1.   yes   no

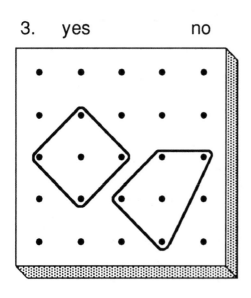

2.   yes   no

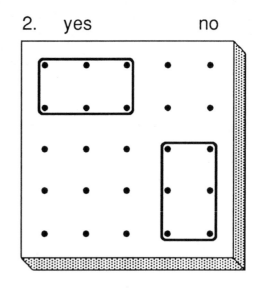

3.   yes   no

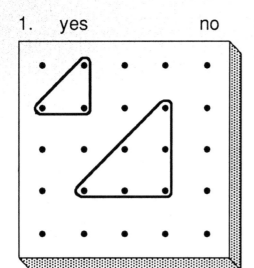

4.   yes   no

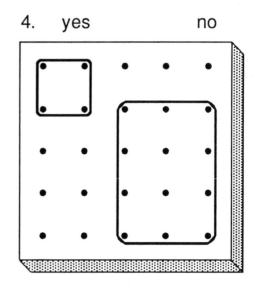

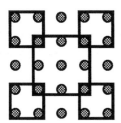

Teacher's Notes

Section B
Perimeter

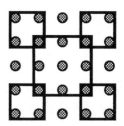

Getting Started

Use a geoboard to mark off one unit with a geoband. You may wish to have students cut a piece of string or straw to be the same length as a unit on the geoboard. Have them stretch a geoband to the length of various units: 3, 4, 2. Next, have students stretch a band to show a diagonal unit. Make sure students see that a diagonal is more than one unit long.

Remind students that the distance around a figure is its perimeter. Then make several figures of different sizes on your geoboard and ask students how they would find the perimeter of each. Have students copy a figure from your geoboard and count the units to find its perimeter. Allow students to create their own figures and find their perimeters. Make sure the students create and measure only closed figures.

Using the Worksheets

Use the worksheets in the order shown at the right.	Units	14
	Making Your Own Shapes	15
	Finding Perimeter	16
	Counting Units	17
	Making Shapes with Different Perimeters	18

As you progress through the worksheets, make sure students are counting and adding the perimeters correctly. Emphasize that sides should not be counted twice. Students may compare figures by wrapping them with string, then stretching the string out and comparing lengths.

Practice

Have students measure other perimeters, such as the distance around books, desks, and windows. Then have them experiment with shapes that have the same perimeter. Have them work in groups to see how many different figures they can make with the same perimeter.

Wrap-up

Ask students to find the perimeters of figures you make on your geoboard. Then have them make figures on their geoboards with perimeters you name.

Name

This is one unit long. •———•
How many units around each shape?

1.
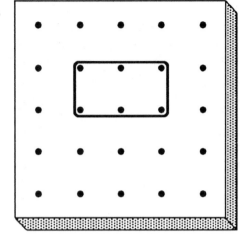

_____ units

2.
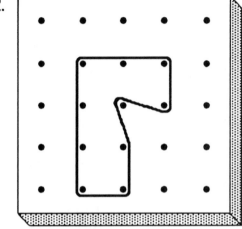

_____ units

3.
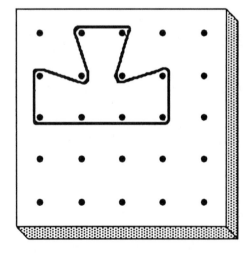

_____ units

4.
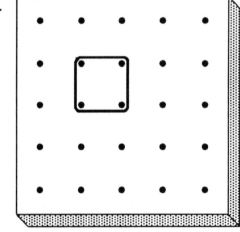

_____ units

Name

Make shapes that measure the number of units shown.

1. 10 units

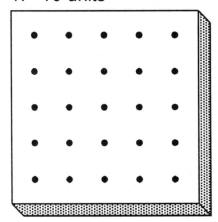

2. 16 units

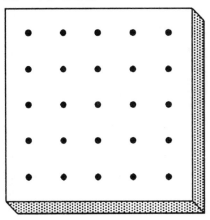

3. 6 units

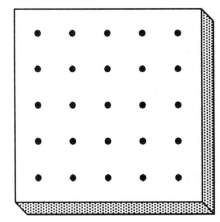

4. 12 units

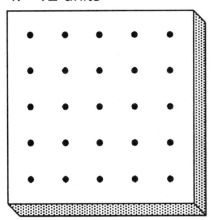

5. Make a shape that looks like this on the geoboard to the right using 8 units.

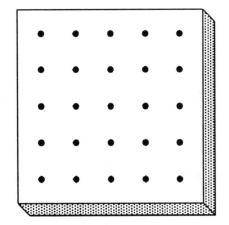

Name

The perimeter of a figure is the distance around it.
Find the units for the perimeter of each shape.

1.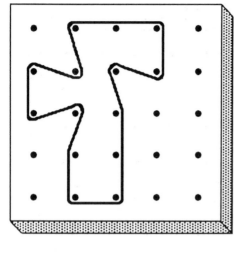

_____ units

2.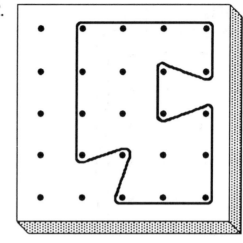

_____ units

3.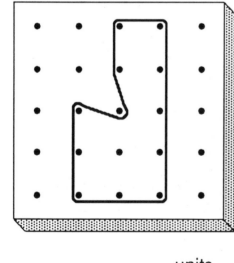

_____ units

4.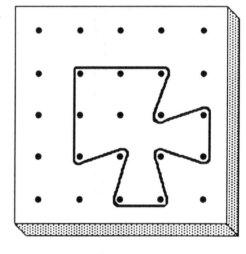

_____ units

5. Which two figures have the same perimeter?   _____ _____

6. On another sheet of paper, draw a different shape with the
   same perimeter as those in Exercise 5.

Find the perimeter of each shape below.

1.
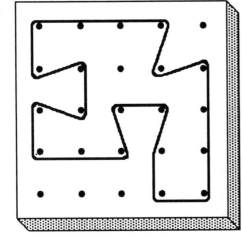

_____ units

2.
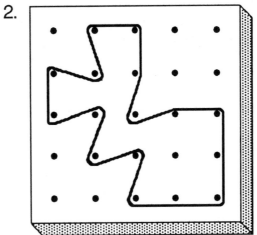

_____ units

3.
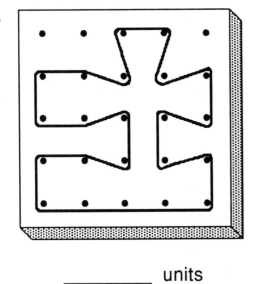

_____ units

4.
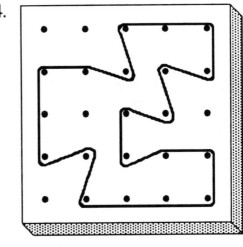

_____ units

5. Which shape above has the greatest perimeter? _____

6. Which two shapes have the same perimeter? _____

7. Which shape has the smallest perimeter? _____

Name

Use your geoboard to solve these problems.

1. Clare is building a dog exercise pen with a perimeter of 10 units. One of the sides will be attached to a wall. What shape would contain the most squares? Draw it on the geoboard below.

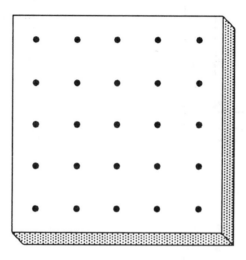

2. Kent drew a figure with 8 squares and a perimeter of 12. Can you make one too? Draw it on the geoboard below.

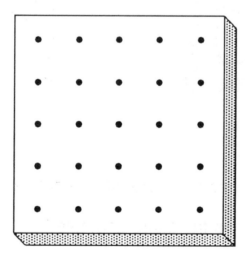

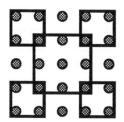

Teacher's Notes

Section C
Area

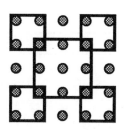

Getting Started

Display one square unit on your geoboard with a geoband. Have students make figures containing three or four unit squares on their geoboards. You may wish to show students a construction paper square you have cut out that is the same size as a square unit on the geoboard. Then make several figures and help students count the unit squares in each figure. Make sure the figures students use are completely enclosed figures. Finally, make a one-half square unit out of construction paper. Use it to measure figures that contain one or more one-half squares.

Using the Worksheets

	Square Units	20
	Counting Square Units	21
	Counting Half-Square Units	22
Use the worksheets	How Many Square Units?	23
in the order shown	Finding Area	24
at the right.	Bisecting Shapes	25
	Finding Bisectors	26
	Lines of Symmetry	27
	How Many Lines of Symmetry?	28

As you progress through the worksheets, make sure students are counting the unit squares correctly; not missing any or counting any twice. Have students cut out shapes the size of a unit square and a one-half unit square to lay on the geoboard to confirm their measurements.

Practice

Have students make figures of a specific number of unit squares. Ask the class to see how many different figures they can make from 3, 5, 7 and 10 unit squares. Some students may want to make separate figures of fewer unit squares, such as 2 or 3, to reach the total. Remind students that the finished figure must be completely enclosed.

Wrap-up

Make a figure on your geoboard. Ask students to first estimate, then count, the number of unit squares in it. Then have students make the same figure on their geoboards. Finally, have them each make a different figure using the same number of unit squares.

This is a square unit. ▢
How many square units
make up each of these shapes?

1.

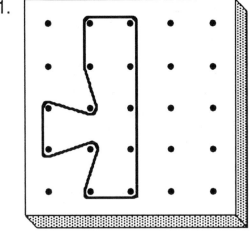

_____ square units

2.

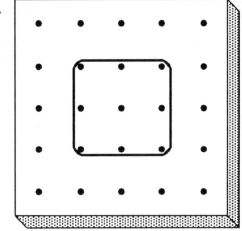

_____ square units

3.

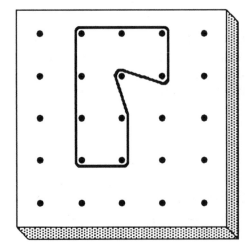

_____ square units

4.

_____ square units

The area of a shape is the number of square units it contains.
Find the area of each figure.

1.

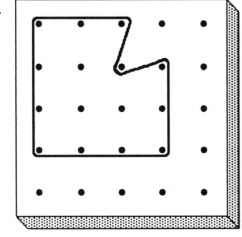

_____ square units

2.

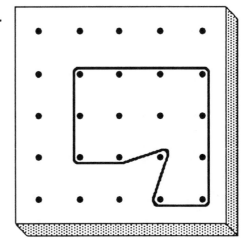

_____ square units

3.

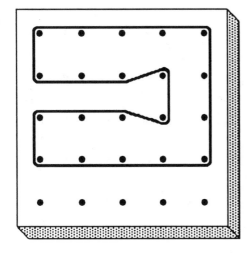

_____ square units

4.

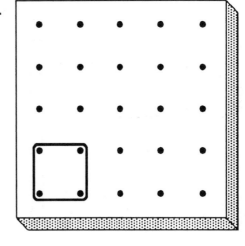

_____ square units

5. On another piece of paper, draw 4 figures with a perimeter of 10 units
   and an area of 4 square units.

Name

This is a one-half square unit.
What is the area of each shape below?

1.

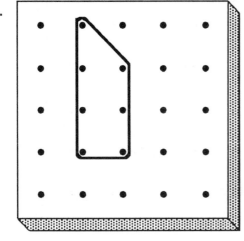

_____ square units

2.

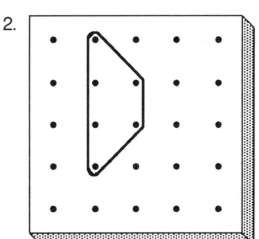

_____ square units

3.

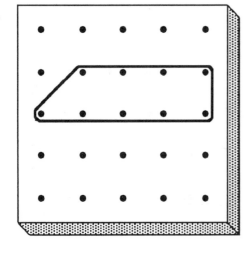

_____ square units

4.

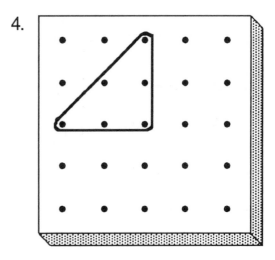

_____ square units

Match each figure with its area in square units.
An area of 1 square unit may look like this.

1.

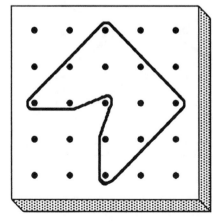

7

2.

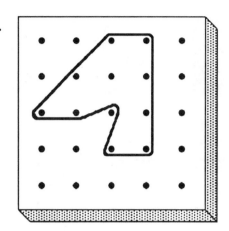

3

3.

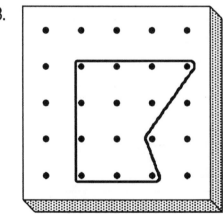

5

2

6

4.

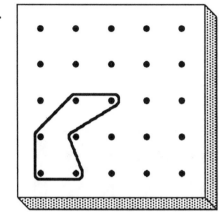

4

5.

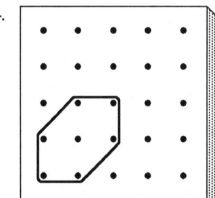

6.

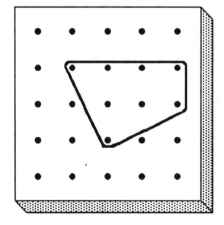

Acme Metal Company manufactured the items below. They are stamped out of a metal plate. Use the geoboards to help find the area of each item.

1.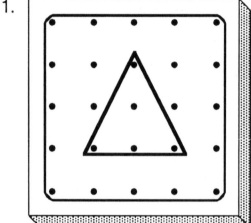

_____ square units

2.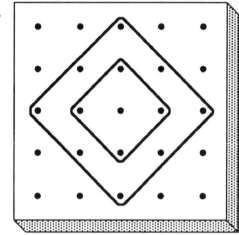

_____ square units

3.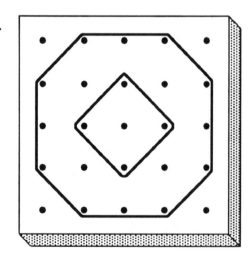

_____ square units

4.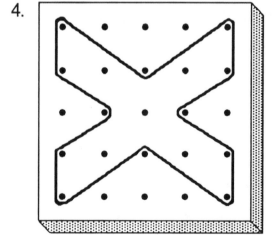

_____ square units

When a shape is bisected, two shapes of equal area are formed.
Copy these shapes and their bisectors on your geoboard.

1.

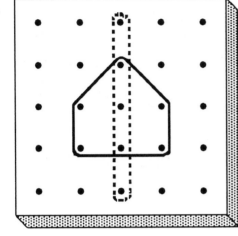

2.

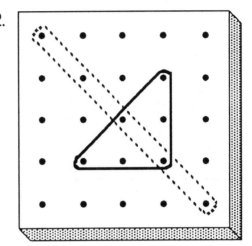

3.

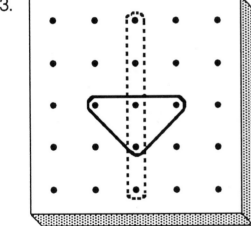

4.

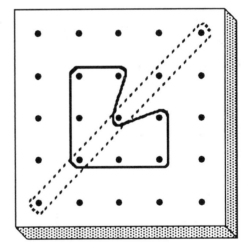

5. On another paper, draw a figure and its bisector.

Name

Find a bisector for each of the shapes below.

1.

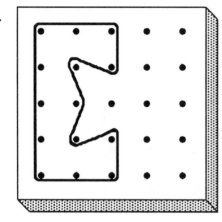

2.

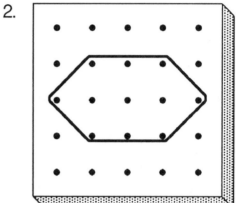

3.

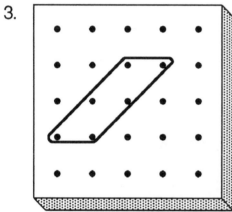

4.

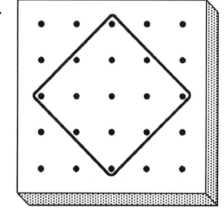

5.

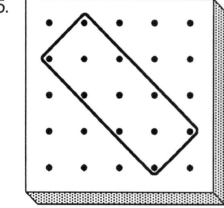

6. Which of the above shapes has more than one bisector?

_____

Copy these shapes and their lines of symmetry which bisect them.

1.
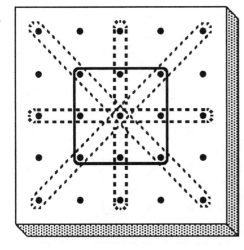

2.
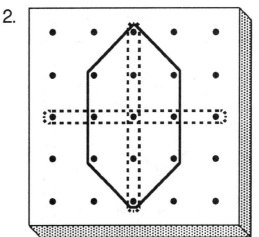

3.
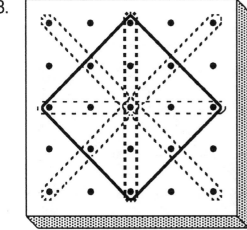

4.
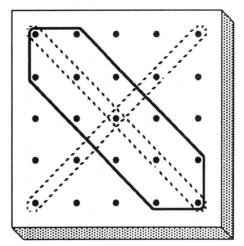

5. On another sheet, draw two lines of symmetry that are at right angles to each other. Draw a figure other than a square or rectangle that has these two lines as bisectors.

How many lines of symmetry may bisect each figure?

1.
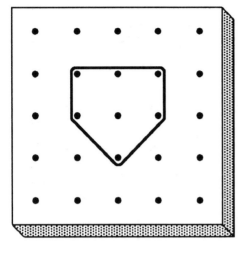

_____ lines

2.
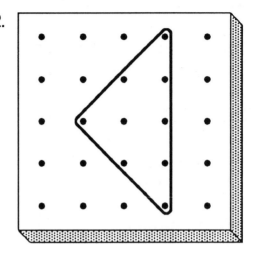

_____ lines

3.
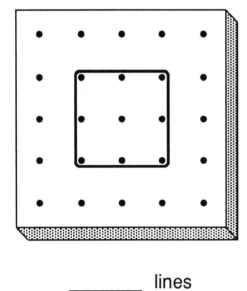

_____ lines

4.
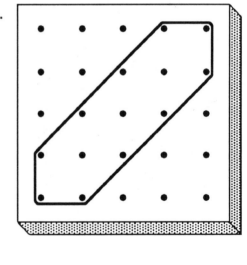

_____ lines

5. On another sheet, make a 6-sided figure that has two lines of symmetry. Make an 8-sided figure that has two lines of symmetry.

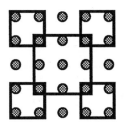 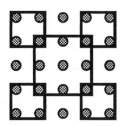

Teacher's Notes

Section D
Polygons

Getting Started   Have students each make a polygon on their geoboards and hold it up. Check to make sure their figures are all polygons. Ask students to count the number of sides in their polygons. Ask students if polygons can have 1 or 2 sides. Talk about why there must be at least 3 sides in a polygon. Ask students who have a 3-sided polygon to hold them up, and have a volunteer name the polygon (triangle). Then have students hold up 4-sided polygons, and ask them to identify this polygon (quadrilateral).

Using the Worksheets

Use the worksheets in the order shown at the right.	Is It a Polygon?	30
	Making Different Polygons	31
	Making Triangles	32
	Making Quadrilaterals	33
	Changing Shapes into Pentagons	34
	Hexagons and Octagons	35
	Naming Polygons	36

As you progress through the worksheets, help students realize that figures with the same number of sides may have different shapes and sizes. Ask students to make a pair of triangles that are similar. Then have them make a pair of triangles that are flips. Ask if these triangles are congruent (Yes.) Have students make a quadrilateral and find its perimeter. Have them hold up quadrilaterals with a perimeter of 4, of 6, and 8 units. Ask why the perimeters of quadrilaterals are always even numbers (The perimeter is 2 times the sum of the length and width.)

Practice   Ask students to make a hexagon and find its area using one-half square units. Warn all students that they may not be able to do this if any side of their hexagon comes between two pegs. Advanced students may be able to find the area of triangle by seeing it is one-half of the area of the rectangle associated with it.

Wrap-up   Have students work in groups of 3. One student chooses a polygon by name, the second student makes the polygon, and the third student checks it. Students exchange roles and repeat the activity until each student has done each task.

Name
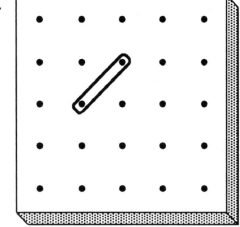

A polygon is a shape with three or more sides. The sides are line segments that meet to form corners. For each shape below circle polygon or not polygon.

1.
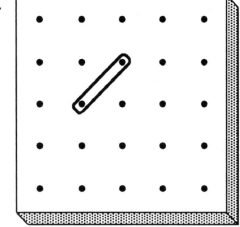

polygon   not polygon

2.
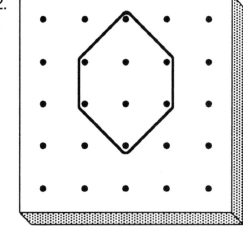

polygon   not polygon

3.
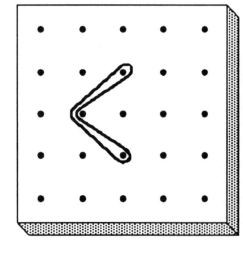

polygon   not polygon

4.
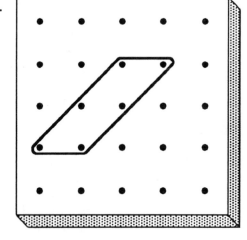

polygon   not polygon

Name

Draw the polygons described below. Then copy them onto your geoboard.

1. four-sided

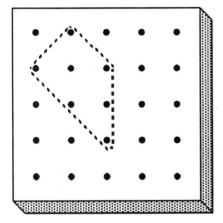

2. six-sided

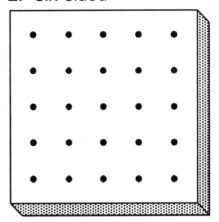

3. three-sided

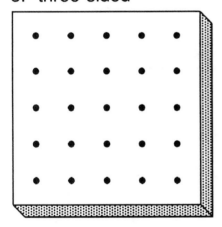

4. five-sided

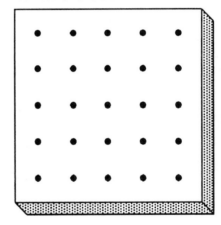

5. seven-sided

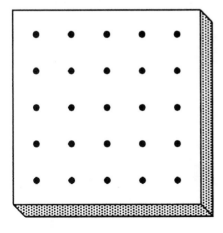

6. eight-sided

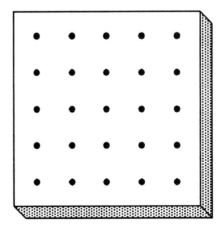

A triangle is a polygon with three sides.
Make triangles that touch the given number of pegs.

1. 4 pegs

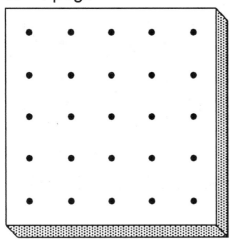

2. 6 pegs

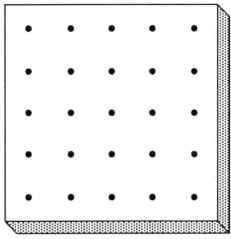

3. 3 pegs

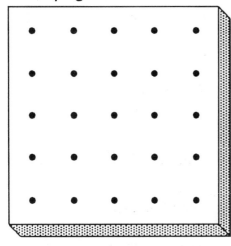

4. 9 pegs

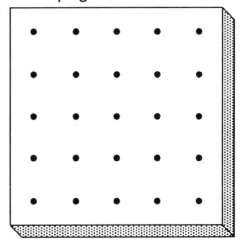

5. Make the largest possible triangle on your geoboard.

6. How many pegs are touched? _____

Name

A quadrilateral is a polygon with four sides.
Make the quadrilaterals that you see below on your geoboard.

1.
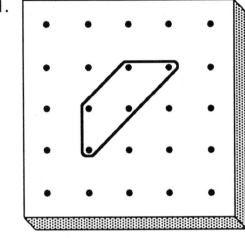

2.
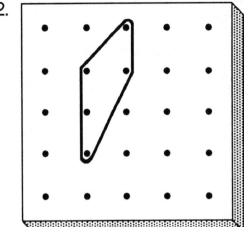

3.
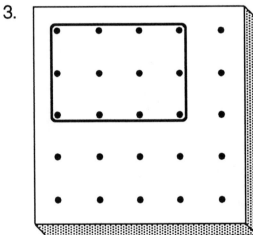

4.
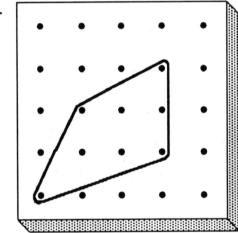

5. Which quadrilaterals above have
   no lines of symmetry that bisect them? _____ _____

A pentagon is a five-sided figure. Change the shapes
you see below into pentagons on your geoboard.

1.
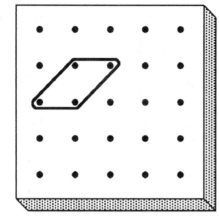

2.
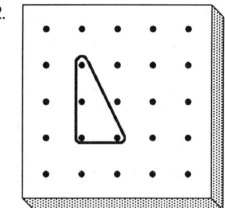

3.
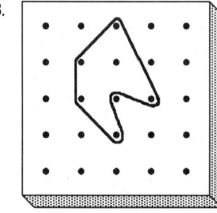

4.
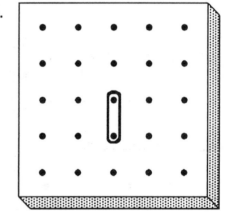

5.
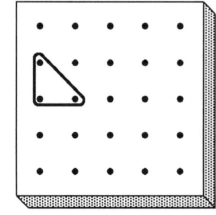

6.
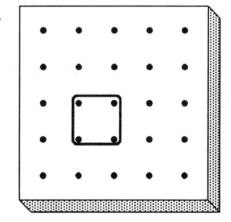

Hexagons have six sides and octagons have eight sides.
Circle hexagon or octagon for each figure.
Then make it on your geoboard.

1.
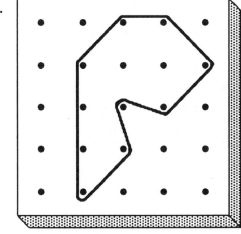

hexagon   octagon

2.
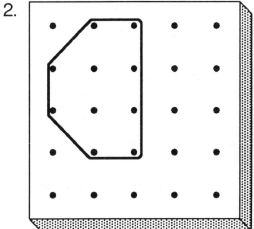

hexagon   octagon

3.
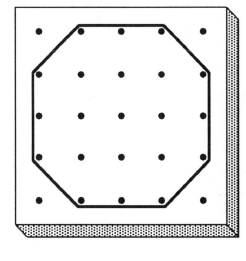

hexagon   octagon

4.
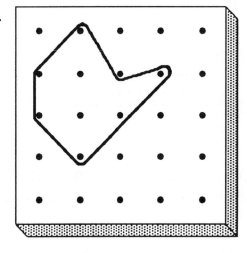

hexagon   octagon

Name

Match each polygon below with its name.

1.
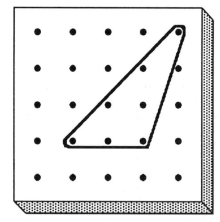

2.
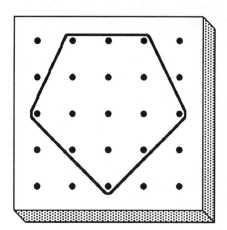

hexagon   pentagon   octagon   triangle

3.
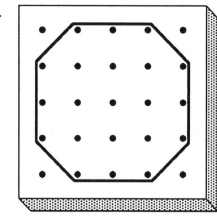

4.
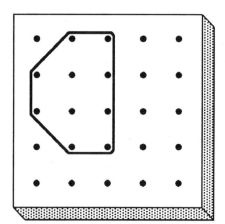

5. On another sheet, draw a pentagon. Use another color to show how you could change its shape to a hexagon.

6. Draw a 7-sided figure.
   Show how you could change its shape to an octagon.

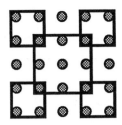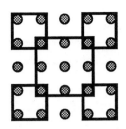

Teacher's Notes

Section E
Quadrilaterals

Getting Started

Tell students to imagine that they are looking at railroad tracks. Ask if the two rails are parallel to each other. (Yes.) Then have students think of a ladder. Ask if the rungs of a ladder are parallel to each other. (Yes.) Ask students to imagine they are holding a rectangular picture frame in their hands. Ask them to touch opposite sides of the frame. Then ask them to touch adjacent sides. Ask which pair of sides is parallel.

Using the Worksheets

Use the worksheets in the order shown at the right.	Parallel Lines	38
	Parallelograms	39
	Trapezoids	40
	Rhombuses	41
	Quadrilaterals and Area	42

As you progress through the worksheets, help students discover that parallelograms can be rectangles, squares, or rhombuses (all with two pairs of parallel sides), but not trapezoids (only one pair of parallel sides).

Practice

Arrange students in pairs. Have each student make a rectangle that is a square and a rectangle that is not a square on their geoboards. Then have them identify these figures on their partner's geoboard. Emphasize that all squares are rectangles, but not all rectangles are squares. Then have each student make a rhombus that is a square and a rhombus that is not a square. Have them identify these figures on their partner's geoboard. Stress that all squares are rhombuses, but not all rhombuses are squares. Ask if a square is always both a rhombus (four equal sides) *and* a rectangle (four equal angles). (Yes.) Have students make charts or diagrams to show how these figures are related: square, rectangle, parallelogram, quadrilateral, trapezoid, and rhombus.

Wrap-up

Ask students to identify and make the following figures: a quadrilateral that has exactly one pair of parallel sides (trapezoid); a quadrilateral that has two pairs of parallel sides (parallelogram); a quadrilateral that has opposite sides parallel and equal with four right angles (rectangle); a parallelogram with all sides equal (rhombus).

Lines that are in the same plane and never meet are called parallel lines.
Copy each line. Then make a parallel line.

1.
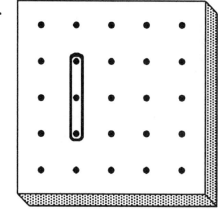

2.
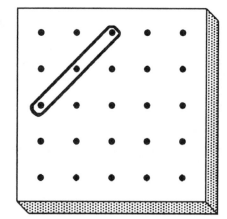

3.
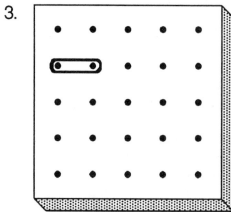

4.
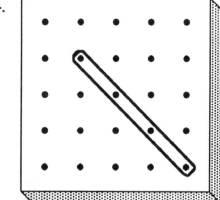

5.
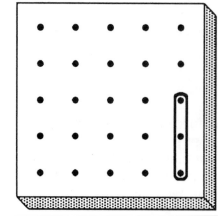

6.
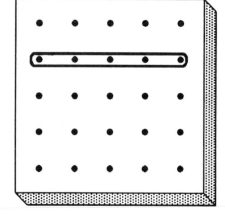

A parallelogram is a quadrilateral whose opposite sides are parallel and equal.

Is each shape below a parallelogram? Circle yes or no.

1. yes   no

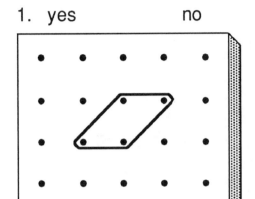

2. yes   no

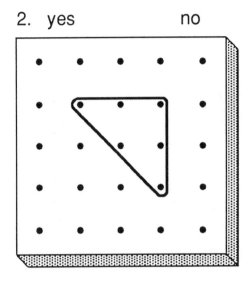

3. yes   no

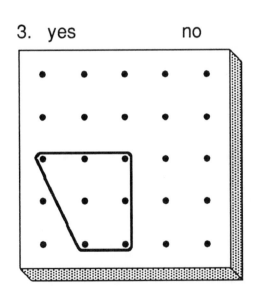

4. yes   no

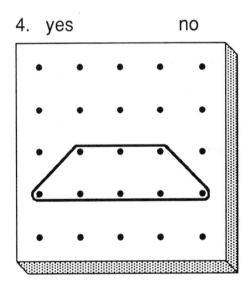

5. Change the non-parallelograms you found above into parallelograms on your geoboard.

6. Is a square a parallelogram? _____

A trapezoid is a quadrilateral with only one pair of sides parallel.

Change the quadrilaterals below into trapezoids on your geoboard.
Then draw the change below.

1.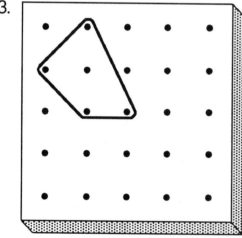

2.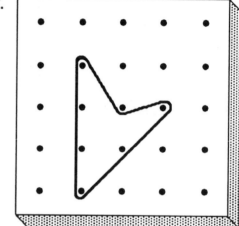

3.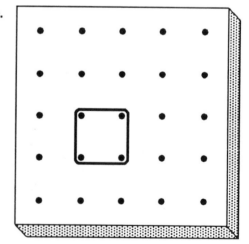

4.

5. On another sheet, make a trapezoid that has a line of symmetry.

Name

A rhombus is a parallelogram with four equal sides. Label each below as parallelogram, rectangle, rhombus, or square. More than one answer may be correct.

1.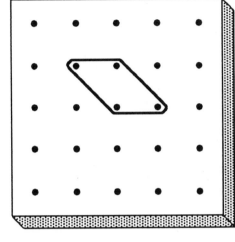

_____

2.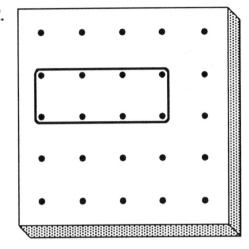

_____

3.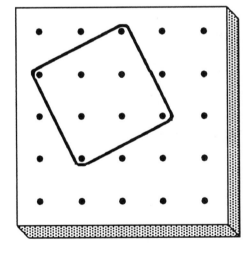

_____

4.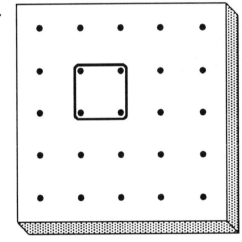

_____

5. For one figure above, write a sentence relating any two of the following words: parallelogram, rectangle, rhombus, square.

_____

_____

Use your geoboard to solve these puzzles.

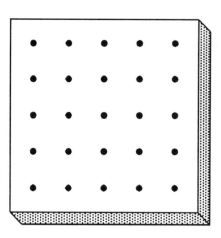

1. Draw a rhombus that has an area of 8 square units.

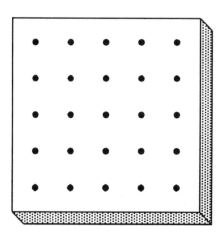

2. Draw a parallelogram with an area of 6 square units, that is not a rectangle.

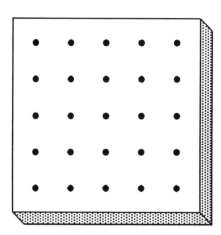

3. Draw a trapezoid with an area of 5 square units.

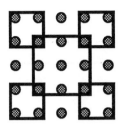

Teacher's Notes

Section F
Fractions

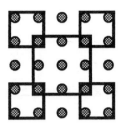

Getting Started

Show students a symmetrical shape on your geoboard. Tell them you will cut the shape in half by dividing the shape with a different colored geoband. Remind students that each section is *one half*. Repeat with several figures that you, and then students, construct. Follow the same procedure to divide some shapes into fourths with two different colored geobands. Call each section *one fourth*.

Next, demonstrate making smaller, identical figures on your geoboard. First make a pair of squares and refer to each one as *one half*. Then make four identical triangles and refer to each one as *one fourth*. Have students make similar figures on their geoboards, first making two identical figures to represent halves and then four figures to represent fourths.

Using the Worksheets

Use the worksheets in the order shown at the right.	Halves	44
	Fourths	45
	Dividing Into Fractions	46
	Identifying Fractions	47
	Displaying Fractions	48

As you progress through the worksheets, remind students how fractional parts can add up to a whole: 2 halves, 4 fourths, 6 sixths, 8 eighths, 16 sixteenths.

Practice

Have students use yellow geobands to make figures that they think can be divided into equal halves and fourths. Instruct them to use a red geoband to divide their figures into halves and add two blue geobands to divide into fourths. Then have students exchange geoboards and trace the figures onto grids. Have students use a green crayon to color in sections at your direction, such as *one-half, one-fourth, three-fourths*.

Wrap-up

Show students figures they can divide into halves and fourths. Have them identify parts such as one-half and three-fourths. Have them identify the end pegs on which the divider geobands will go.

Name

Make 2 equal parts. Color one-half.

1.
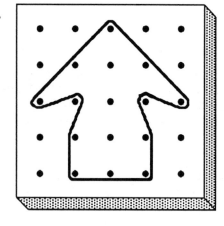

2.
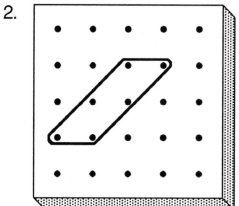

3.
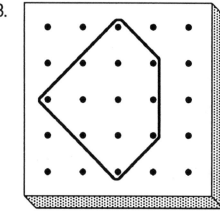

4.
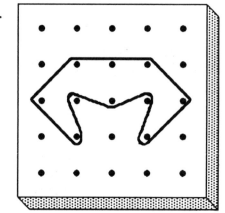

5.
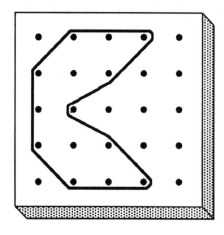

6.
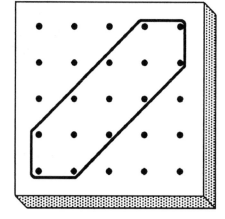

Name

The shapes are divided into 4 parts. Shade the given amount.

1. three-fourths

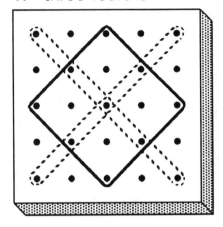

2. one-fourth

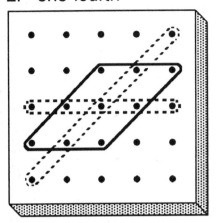

3. two-fourths

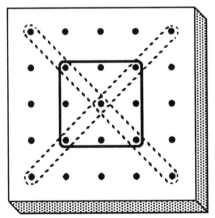

4. four-fourths

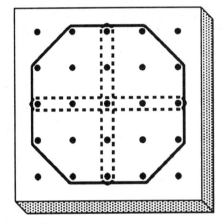

5. one-half

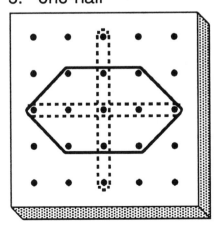

6. three-fourths

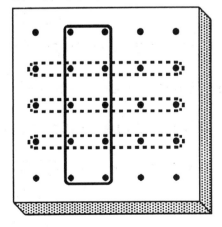

Think of each geoboard below as a square with sides of five units. Divide each into equal parts as shown.

1. Divide into fourths.

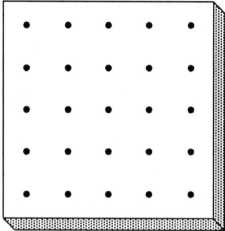

2. Divide into halves.

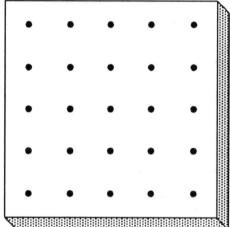

3. Divide into eighths.

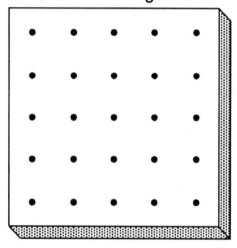

4. Divide into sixteenths.

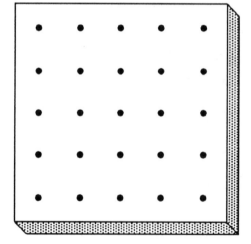

5. Make a new shape on your geoboard.

6. Divide your shape in half.

7. Now divide your shape into fourths.

Are the correct number of shapes shaded? Circle yes or no.

1. one-fourth

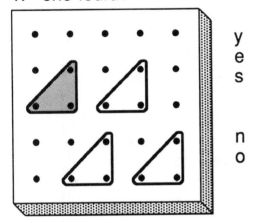

y
e
s

n
o

2. one-half

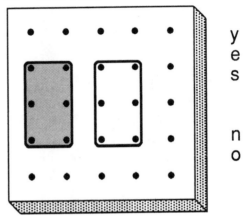

y
e
s

n
o

3. three-fourths

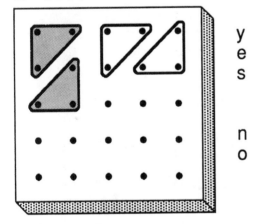

y
e
s

n
o

4. four-fourths

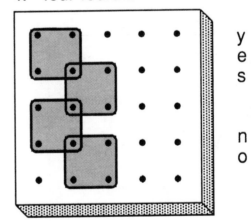

y
e
s

n
o

5. two-eighths

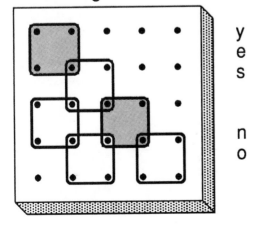

y
e
s

n
o

6. three-sixths

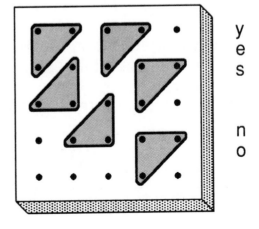

y
e
s

n
o

Shade the correct number of shapes.

1. one-fourth

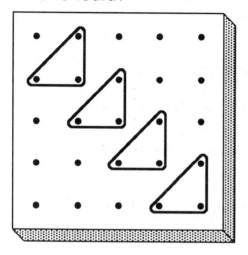

2. four-fourths

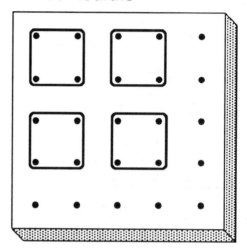

3. two-fourths

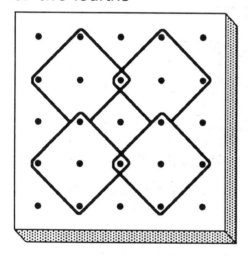

4. three-fourths

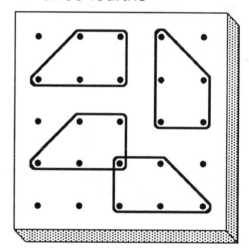

5. On another sheet, for each group above, write the fractional part that is not shaded.

6. Write a sentence relating one-half and two-fourths.

_____

_____

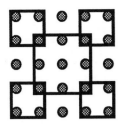

Teacher's Notes

Section G
Rotation and Design

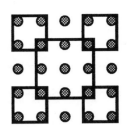

Getting Started
Hold your geoboard in front of you as you make a simple figure such as a square or triangle to rotate. Point to the peg or pegs and explain that they will be the *rotation* points. Demonstrate a one-half turn with a different colored geoband. Show students how the rotated figure is the same shape and size as the original figure. Repeat until students volunteer to make the one-half turns. Change the position of the original figure from the left to the right, then move it up and down.

Repeat the procedure above with one-quarter turns. Show students how two successive one-quarter turns will result in a one-half turn. Eventually use more complicated figures for students to rotate.

Using the Worksheets

Use the worksheets in the order shown at the right.	Copying One-Quarter Turn	50
	Rotating One-Quarter Turn	51
	Copying One-Half Turn	52
	Rotating One-Half Turn	53
	Rotating Figures	54
	Making Designs With Rotation	55
	Designing Objects	56

As you progress through the worksheets, first provide the initial figure for students to mirror. Then have students copy the original figure and its rotation, and finally have them make the rotation independently.

Practice
Have students plan rotations of their own. One student makes a figure and passes it to another student who makes the turn. A third student identifies the rotation as a one-half or one-quarter turn.

Next, have students stay in their groups to plan a design for a geoboard, using as many shapes and different colors as they can. Award prizes for the most original design, the best use of rotation, and so forth.

Wrap-up
Ask students to make a figure. Then ask them to make a design that completes a four-fourths rotation using four different geobands.

Name

The dotted shape has been rotated one quarter-turn. Copy each pair of shapes.

1.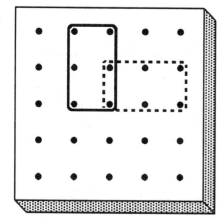

2.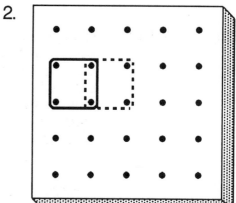

3.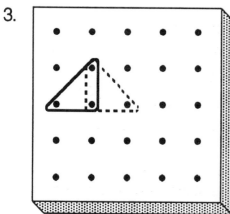

4.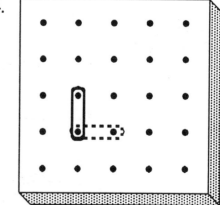

5.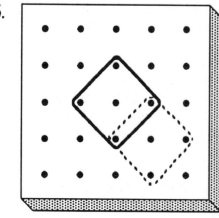

6.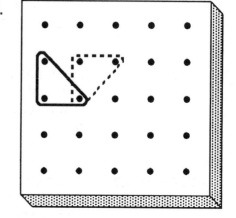

Turn each shape one-quarter turn.

1.

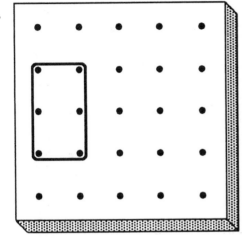

2.

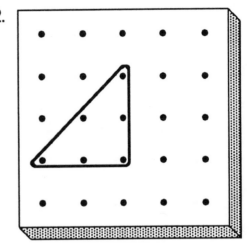

3.

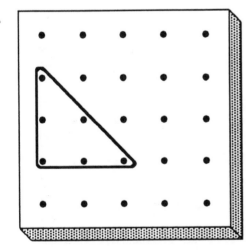

4.

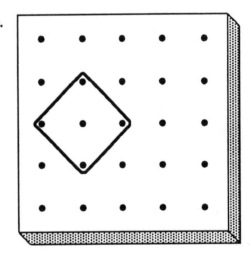

5. Think of a spinner with 4 equal parts. The arrow is on a line. It is a "liner." Write about the number of one-quarter turns the spinner can move to show other liner positions.

The dotted shapes have been rotated one-half
turn. Copy each pair of shapes.

1.
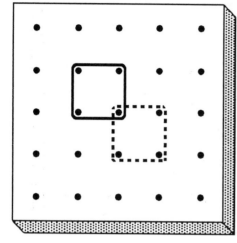

2.
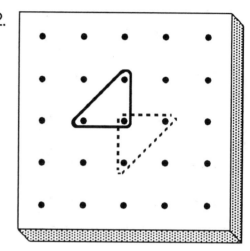

3.
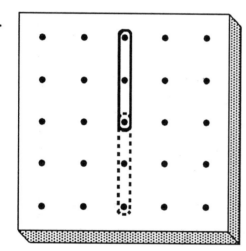

4.
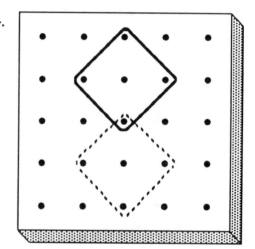

5. Figure 4 shows one rotation of one-half turn. From the same
   starting position, show three other possible one-half turns.
   Use another sheet.

Turn each shape one-half turn.

1.

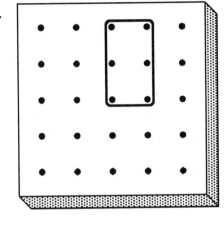

2.

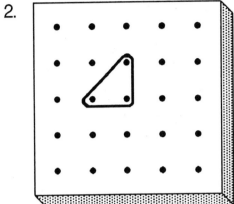

3.

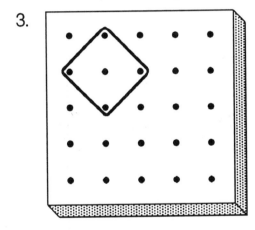

4.

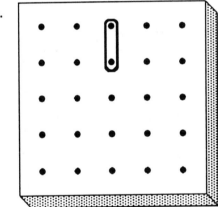

5.

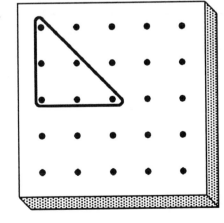

6.

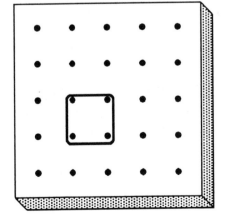

Rotate each figure.

1. one-half turn

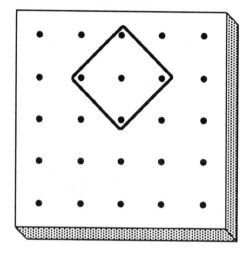

2. one-quarter turn

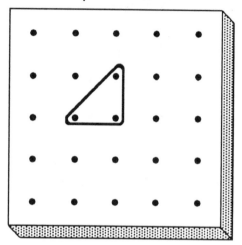

3. one-quarter turn

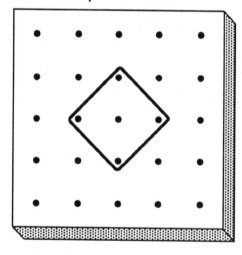

4. one-half turn

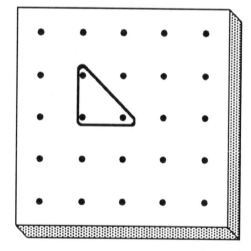

5. On a separate sheet, make a square.
    Make a one-quarter turn and then another one-quarter turn.
    Are two one quarter turns the same as one half turn? _____

Make a design by rotating the given shape.

1.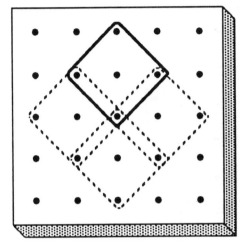

2.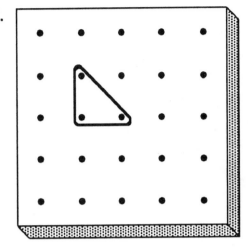

3.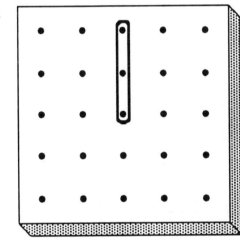

4.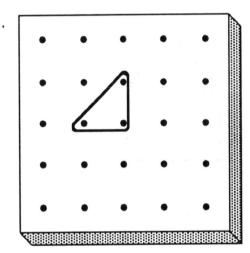

5. On another sheet, try making designs using other shapes and rotations.

Name

Design a figure on your geoboard that looks like each of the following.

1.

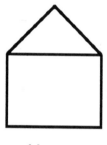
STOP

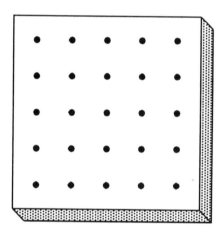

2.

House

3.

U
S
A
Rocket

Name

1.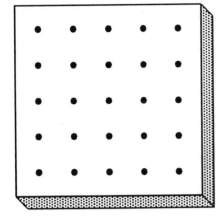

2.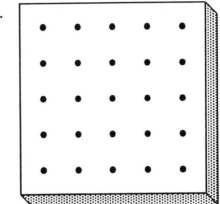

3.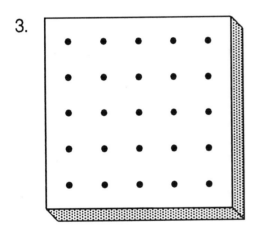

4.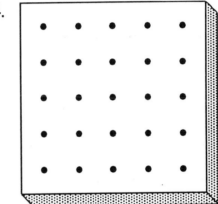

5.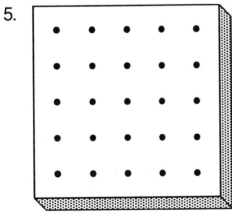

6.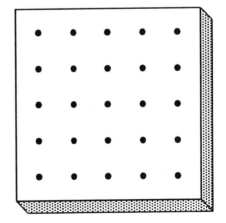

Name